MW00592388

Ferrari

THE EARLY BERLINETTAS & COMPETITION COUPES

By Dean Batchelor

Motorbooks International
Publishers & Wholesalers ®

To Warren Fitzgerald: Writer, artist, photographer, historian, automotive enthusiast and, most of all, a friend. We miss you.

This edition first published in 1993 by Motorbooks International Publishers & Wholesalers, PO Box 2, 729 Prospect Avenue, Osceola, WI 54020 USA

© Dean Batchelor Publications, 1974, 1993

This is a reissue of the 1974 original edition with corrections and revisions

All rights reserved. With the exception of quoting brief passages for the purposes of review no part of this publication may be reproduced without prior written permission from the Publisher

Motorbooks International is a certified trademark, registered with the United States Patent Office

The information in this book is true and complete to the best of our knowledge. All recommendations are made without any guarantee on the part of the author or Publisher, who also disclaim any liability incurred in connection with the use of this data or specific details

We recognize that some words, model names and designations, for example, mentioned herein are the property of the trademark holder. We use them for identification purposes only. This is not an official publication

Motorbooks International books are also available at discounts in bulk quantity for industrial or sales-promotional use. For details write to Special Sales Manager at the Publisher's address

Library of Congress Cataloging-in-Publication Data
Batchelor, Dean.
    Ferrari: the early Berlinettas & competition coupes / Dean Batchelor.
      p. cm.
    Originally published: Incline Village, Nev.: DB Publications, 1974.
    ISBN 0-87938-706-8
    1. Ferrari automobile—History.   2. Automobiles, Racing—History.   I. Title.
TL215.F47B37   1993
629.228—dc20   92-33866

Printed and bound in the United States of America

FERRARI The Early Berlinettas & Competition Coupes

CONTENTS

FORWARD

BERLINETTA, translated literally from the Italian, means "little sedan" (a true sedan being a *Berlina).* In U.S. terminology, a berlinetta would probably be a club coupe, or club sedan, as exemplified by the General Motors "sedanette" bodies on some 1941–1952 Chevrolet, Buick, Pontiac, Olds and Cadillac cars. These were fast-back designs and while most were two door models, there were also four door versions. All were 5–6 passenger bodies—in effect, little sedans.

The Italian use of this terminology invariably refers to a two-passenger coupe usually, but not always, built for (or at least capable of) competition in the appropriate class. The word berlinetta *almost* always is used in conjunction with a fast-back roof design.

Fast-back, or slope-back, styling is more popular than it is practical inasmuch as it is not always aerodynamically sound, and in *gran turismo* versions usually severely limits luggage space and/or passenger access compared to notch-back designs. It is, however, more appealing visually and automotive designers and stylists the world over have made the most of this aspect.

Pinin Farina pioneered the look in his Fiat-based Cisitalia of 1947; a car that put both Carrozzeria Pininfarina and fast-back design in the automotive world limelight.

The first Ferrari to carry what was going to become *the* look for high performance gran turismo cars were built by Carrozzeria Touring and Alfredo Vignale in 1950. Later designs and variations of what came to be associated with the term *berlinetta* were built by Pininfarina, Zagato, Stabilimenti Farina, Bertone, Scaglietti, Ghia, Ghia-Aigle, Boano, Drogo, and Neri & Bonacini.

Enzo Ferrari's factory does not normally build bodies for its chassis, preferring to remain involved with engineering design and construction, rather than bodywork. Ferrari relies on outside coachbuilders for the bodies of his road cars; however he has always retained complete and absolute approval for all designs if they were to be built for Sefac Ferrari—either for racing or for normal road use.

Many chassis have been purchased by, or loaned to, coachbuilders for special show cars, and in recent years some older production Ferraris have been rebodied by amateur body designer/builders or by established smaller coachbuilders for customers.

In this book we show only the earlier Ferraris with berlinetta type coachwork from the first examples, built by Carrozzeria Touring on 166, 195 and 212 chassis, through the Pininfarina-designed, Scaglietti-built 250 GT short wheelbase berlinetta of 1959–62. I make no claim to the inclusion of all berlinettas built during this period. It would be impossible. Instead, I will show a representative selection of berlinetta body/engine combinations through the first 12 years of Ferrari construction. Subsequent books will cover Ferrari "spyder" and competition roadsters, and later berlinettas.

Not all the cars covered in this book were given the berlinetta designation either by the factory or the coachbuilder. The first cars, for example, were called by their chassis designations; 166 Le Mans, 166 Inter, 195 Sport, 250 Europa, 340 America, etc. In most cases the factory referred to the cars as simply GT (for Gran Turismo, or grand touring), or *coupe competizione.*

All the cars in this book, however, fit our self-established definition of a gran turismo berlinetta and, furthermore, were significant cars in both the history of Ferrari and the coachbuilders' art.

These were, and are, some of the most exciting cars ever seen on the road or on the track, and in most cases they were at home in both places.

Many early Ferraris did not handle particularly well. Some models had weak rear axles. Their complexity has made mechanics weep. Some had heating problems. Plug fouling was a problem if driven in traffic. Cooling could also be a problem in stop and go driving. There was no provision for front end alignment other than toe-in or toe-out adjustment (camber and caster were fixed at the factory and that was it!). Clutches, even on competition models, were notoriously marginal. Yet, the visual and audible excitement of these early 12 cylinder Ferraris at a time when other competition cars were fours, sixes or V-8s, set the red cars from Maranello apart from anything else on the track. Win, lose or draw, Ferraris made the race for the spectators.

It is with a great deal of personal pleasure and admitted love for a period past—one we shall never again witness—that this first book in a series is presented. I hope you get as much enjoyment reading, and looking at, FERRARI The Early Berlinettas & Competition Coupes, as I had in compiling it.

Dean Batchelor

166 Mille Miglia

The year 1946 will go down in history as one of the most significant in the annals of automobile racing. In Maranello, Italy, a new 1500cc V-12 engine had been designed by Gioacchino Colombo and was being tested and developed by Colombo, and Luigi Bazzi, for Enzo Ferrari.

The Italian automotive magazine *Inter Auto* published a description of this new engine in its November–December 1946 issue, and noted that it had been designed as a high performance sports car engine which could be tuned to competition capabilities, and be supercharged for Grand Prix use.

If the message was lost on the racing world, it shouldn't have been. Ferrari had been the official Alfa Romeo team before World War II, and Colombo had been the protege of Vittorio Jano who was responsible for the P2 and P3 Alfa Romeos, among others. Colombo and Ferrari had worked together since 1929, and they knew how to design, build, prepare and field a racing car—or a team of cars.

The first cars built carrying the Ferrari name were raced at Piacenza, Italy, on May 11, 1947. Franco Cortese drove a 125S and led the race until the last lap when the fuel pump seized (the early Ferraris had aircraft-type centrifugal pumps). Vindication came quickly as Cortese won two weeks later at Rome's Caracalla circuit; again at Vercelli on June 1, at Vigevano on June 15, and at Varese on June 29. Tazio Nuvolari then won the sports class in two more races for Ferrari in July.

By August, the engines of two of the original three cars had been bored to increase their displacement from 1500 to 1908cc and were now called Tipo 159s. At Turin on October 12, 1947 Raymond

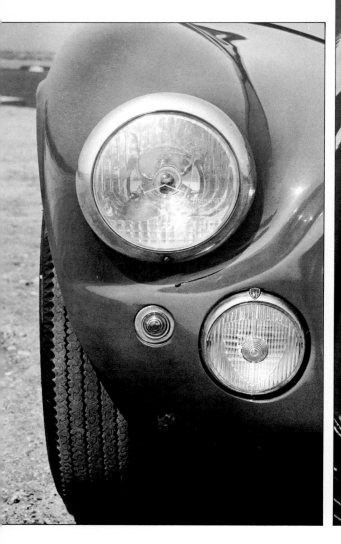

Sommer won 1st overall with a 159.

On April 3, 1948 a still larger version, called the 166—with 1995cc—won the Targa Florio, driven by Clemente Biondetti and Count Igor Troubetzkoy. This car was then fitted with a berlinetta top and with it Biondetti won the Mille Miglia. This is believed to be the first closed Ferrari ever built, and the first car to win the Mille Miglia carrying the name Ferrari. But it wouldn't be the last. Ferraris won the Mille Miglia eight times outright before the race was cancelled in 1957.

Ferrari began production of 166 chassis/engines, equipped with a single Weber carburetor and called the Inter, for street use, in 1948. Stabilimenti Farina (not to be confused with Pininfarina) built both open and closed bodies on this version, as did Ghia, and Carrozzeria Touring built some rather awkward looking coupe bodies for the 166 Inter.

The 1949 Ferrari yearbook carries an advertisement that had appeared in Italian magazines showing a photo of a 166 Inter coupe, and drawings of a 166 MM Barchetta and the 166 Le Mans (all with bodies by Touring), the latter type being the subject of this feature.

The Type 166 engine was offered, in various versions, until 1953, even though bigger engines were available at the same time. Inasmuch as the 195 and 212 were bored-out versions of the 166, and therefore had the same external dimensions and weight, the same chassis could be used for all three cars.

Early Ferraris, other than the single carburetor touring versions, were really dual-purpose cars in the truest sense of the word. They competed successfully in the Mille Miglia, Targa Florio, Le Mans, Tour de France and the occasional international rally, and it was a competition car that could also be driven to and from the event if the owner chose to do so.

Judging from the aforementioned advertisement, one could surmise that Ferrari intended the open version to be raced in the

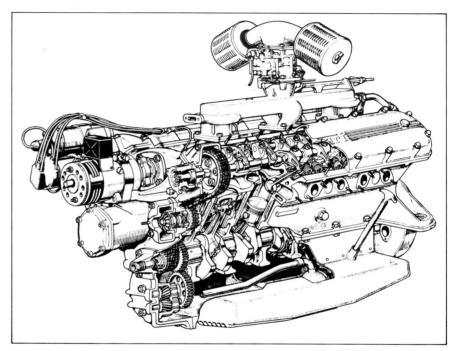

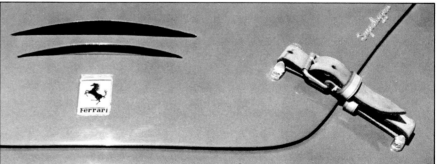

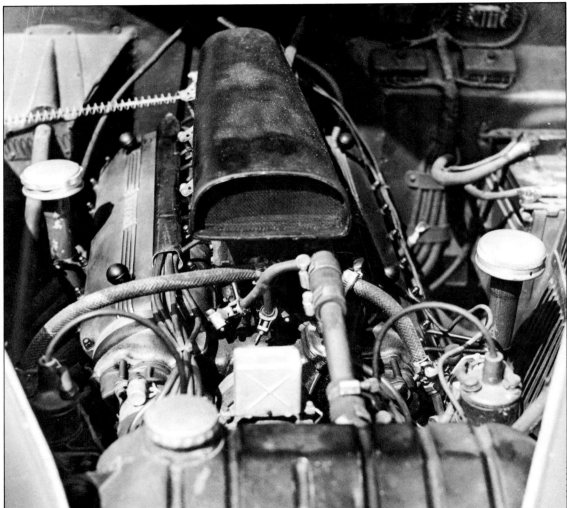

Body of 195 is same size and shape as 166 but has minor detail differences—most noticeable being the grill, externally, and the carburetor air cleaners.

Mille Miglia, and the closed coupe, such as we see on these pages, at Le Mans. Somehow, it didn't work out that way and the Touring-built coupe came to be known as a Mille Miglia too, probably in honor of Biondetti's victory in 1948.

The heart of the little coupe, as in all Ferraris, is the engine. The Colombo design is a 60 degree V-12 with the cylinder block, heads, and sump cast of aluminum alloy. The cylinder liners were cast-iron, shrunk into place and further positioned by a shoulder about two thirds of the way down the liner which mated with the block for a metal-to-metal joint, and a shoulder at the top which was compressed down by tightening of the cylinder head. Compression and water passage were sealed at the block-head joint by individual copper-asbestos gaskets around each hole.

There was a 20mm offset between cylinder banks so the H-section rods could run side by side on the crank throw. The rods were split at 40 degrees from the rod center line so they could be removed from the top of the engine, and all Colombo-designed engines used this type of rod until the 250 Testa Rossa in 1958.

In spite of Colombo's genius, these new engines did give some trouble, although most of the problems were cured by later Ferrari engineers Bazzi, Lampredi, Massimino, Jano and Rocchi.

Among the problems were abnormal wear on the finger-type cam followers, siamesed intake ports which made carburetor tuning difficult, a triangular stud pattern around the bores that made it difficult to torque the heads down effectively. The long dimension from the top of the barrel to its lower positioning-shoulder allowed the differential expansion characteristics (of the cast-iron sleeve and the aluminum block) to cause loss of compression at the copper O rings when the engines got hot during competition.

The frame design for both roadster and coupe was the same; deep oval section steel main tubes with a box member at the front and an X member in the center—an assembly often referred to as a "ladder" type frame.

At the front, suspension was independent with double A-arms on each side and a transverse leaf spring. This system was used in all Ferrari road cars until 1955 when coil springs were used on the second series Europas. The very early Ferraris also had the lever-type shock absorbers built in unit with the front upper A-arm, but later cars carried the shock as a separate unit. Rear suspension on the early cars was a live axle with semi-elliptic springs and Houdaille lever shocks.

Carrozzeria Touring bodywork carried the patented name *Superleggera* (super light), and it was unique to this builder—at least for production or semi-production cars. A skeleton of small diameter steel tubing—anywhere from ¼ to ½ inch in diameter—was made for the general body shape and for hood, door, wheel and "trunk" openings. To this framework was attached aluminum body panels which were rolled over the tubing to secure the assembly together. The result was a lightweight, rigid body structure, but one which contributed little if any stiffness to the car as a whole. But inasmuch as Ferrari always used separate frame and body construction anyway, the body wasn't required to do any more than cover the assembly of parts that go to make up a car.

As was standard practice in the early Fifties, the steering was on the right side. I've asked countless "authorities" or "experts" about this—particularly in view of the fact that driving in Italy is done on the right side of the road. The only sensible reply I've received is that, in Europe, races are run clockwise on the circuit and the driver should be on the inside of the majority of curves, and on the pit side of the car when it stops at the pits.

This makes a certain amount of sense although most drivers I've talked to don't think it makes much difference which side of the car they're on. Regardless, Ferrari still puts his racing driver on the right (as do most current builders of sports racing and prototype cars) but the driver of his customer-oriented road car is now on the

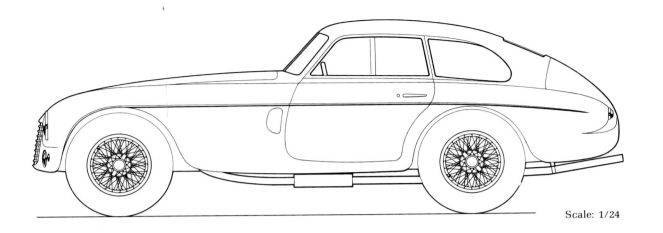

Scale: 1/24

166 Mille Miglia

ENGINE

Type ....Colombo-designed, water-cooled, 60 degree V-12
Bore/stroke, mm/inches 60.0/58.8, 2.362/2.315
Displacement, cc/cubic inches 1995/122
Valve operation: Single overhead camshaft on each bank with finger
    followers & rocker arms to inclined valves
Compression ratio...8.5:1
Carburetion .... Three Weber 36 DCF twin-choke, downdraft
BHP (Mfg)..125 @ 7000 rpm

DRIVE TRAIN

Clutch ....Single dry plate
Transmission: Five speed, non-synchromesh, direct drive in 4th gear
Rear axle ....Live
Axle ratio .... Optional—4.66 or 5.0:1

CHASSIS

Frame................................Welded tubular steel, ladder-type
Wheelbase, mm/inches 2250/88.6
Track, front, mm/inches 1270/49.8
    rear, mm/inches 1250/49.2
Front suspension: Unequal length A-arms, transverse leaf spring, anti-
    roll bar
Rear suspension: Live axle, semi-elliptic springs. Rear shock arm &
    triangular bracket for axle location
Shock absorbers .... Houdaille hydraulic lever-action
Brakes .... Hydraulic, aluminum drums with iron liners
Tire size .... 5.90 x 15
Wheels..Borrani wire, center-lock knock-off

GENERAL

Length overall, inches 146.0
Width 60.0
Height 46.0
Curb weight 2125
Body builder .... Carrozzeria Touring

left except for cars built for export to England or Japan.

Unless you live in one of the countries that drives on the left side of the road, owning a classic Ferrari with right hand drive can be a test of nerves and disposition. I've always felt safer with a passenger along who could check oncoming traffic to tell me if and when it was safe to pass, otherwise you have two-thirds of the car on the wrong side of the road before you know for sure.

Frustrating to drive or not, these diminutive Touring-bodied Ferraris are among the most desirable of classic Ferraris (or any classic car, for that matter) and have become genuine collector's items. Most of the existing 166, 195 and 212 Ferraris have been, or are being, restored and well they should be. They are the

epitome of early Ferraris.

The Touring-bodied 195 S pictured on these pages was featured in the March, 1965 issue of *Road & Track*. After the article was published, the magazine received a letter from Jim Kimberly commenting on the car: "In 1950 when we raced in Buenos Aires, Briggs Cunningham shipped the car, but was unable to get away himself and asked me to drive it for him. Of course it didn't take much wrist twisting and I jumped at the chance. It was a little small for me and over the rough section, and I do mean rough, on the back side of the course, I really appreciated a crash helmet for the first time. Without it, I am sure I would have flattened out my skull completely."

195 Inter

The pressure of competition had caused Ferrari to enlarge the original Colombo-designed 1500cc V-12 engine to 2000cc, and in 1950 it was bored still further to 65mm (retaining the 58.8mm stroke) for a capacity of 2341cc.

This new model was called the 195 and came in two versions; the Inter, with a single Weber 32DCF twin-choke carburetor, was rated at 130 bhp at 6000 rpm, and the Sport, with three 32 DCF Webers, produced 145 bhp at 6600 rpm.

The Inter was built as a road touring car but the Sport was intended for competition and the extra horsepower came from a combination of the two additional carburetors plus a higher compression ratio—8.5 for the Sport and 7.5:1 for the Inter.

Very few 195s, in either Inter or Sport versions, were made and few of them achieved much success in racing as the larger 212 Inter and Export were introduced in late 1950. A Touring-bodied 195 did win the 1950 Mille Miglia, in the hands of Gianinno Marzotto, at 76.79 mph. The race was run in rain, mist and, sometimes, hail at various times on the 1000-mile circuit of Italy.

The same 195 coupe was entered in the 24 hours of Le Mans in June 1950 and, driven by Raymond Sommer, ran away from the field only to make a lengthy pit stop to repair a broken generator mount. The repairs were made, but were not totally successful and the car later retired when the repairs didn't last.

The frame, suspension and brakes of the 195 were basically the same as that of the 166 but the Inter wheelbase was lengthened—from 2200 to 2500mm (88.6 to 98.5 inches). Track was the same as the 166 MM at 1270mm front and 1250mm rear.

In view of the shorter wheelbase of the Touring-bodied 195s it is

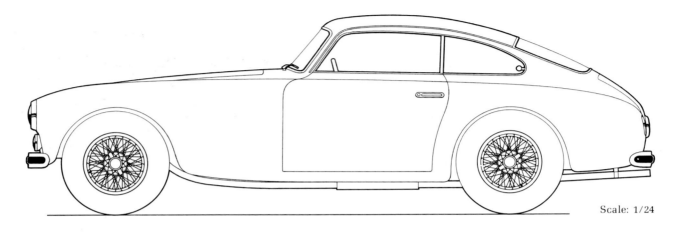

Scale: 1/24

195 Inter

ENGINE

Type ....Colombo-designed, water-cooled, 60 degree V-12
Bore/stroke, mm/inches 65.0/58.8, 2.562/2.315
Displacement, cc/cubic inches 2340/143
Valve operation: Single overhead camshaft on each bank with finger
  followers & rocker arms to inclined valves
Compression ratio...7.5:1
Carburetion ....Single Weber 32 DCF twin-choke, downdraft
BHP (Mfg)...130 @ 6000 rpm

DRIVE TRAIN

Clutch ....Single dry plate
Transmission: Five speed, non-synchromesh, direct drive in 4th gear
Rear axle...Live
Axle ratio 4.66:1

CHASSIS

Frame...Welded tubular steel, ladder-type
Wheelbase, mm/inches 2500/98.4
Track, front, mm/inches 1270/49.8
  rear, mm/inches 1250/49.2
Front suspension.........Unequal length A-arms, transverse leaf spring
Rear suspension...................................Live axle, semi-elliptic springs
Shock absorbers ....Houdaille hydraulic lever-action
Brakes ....Hydraulic, aluminum drums with iron liners
Tire size, front/rear ....5.90 x 15
Wheels.......................................Borrani wire, center-lock knock-off

GENERAL

Length overall, inches 156.0
Width 60.0
Height 47.0
Curb weight 2100
Body builder ....Alfredo Vignale

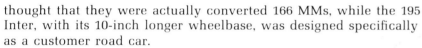

thought that they were actually converted 166 MMs, while the 195 Inter, with its 10-inch longer wheelbase, was designed specifically as a customer road car.

Bodies for the 195 S were built by Carrozzeria Touring, and for the 195 Inter they were generally built by Ghia or Vignale. Like all early Ferrari production, however, there were variations and Vignale apparently built several bodies for the 195 S.

The car featured on these pages is a further anomaly as its serial number (115 S) indicates it should be a "Sport" yet it has the single carburetor engine and longer wheelbase of the 195 Inter. Students of Ferrari lore find this frustrating but not surprising, as the early years were so experimentive for Ferrari that few if any cars were the same—even in the same series.

The Vignale-bodied 195 shown is a graceful and pleasant design if not one of the most outstanding of Vignale's efforts. Being a 1950 model, it preceded both the Vignale cross-bar and the later standard Ferrari egg-crate grille motifs. The grille is not too busy, as were some later versions of this style, and exemplifies the still experimental designs of Vignale. The inclusion of two driving lights, two parking lights, and several club badges tends to be distracting, however.

Nevertheless, this 195 has to be called a handsome car, and one of the great things about owning an older Ferrari is that if the design was good in the first place, it still holds up well today. The suspension and running gear are definitely vintage, and compared to new designs are obviously inferior, but the owner can take pride in enjoying one of the best cars of its era, plus a body style that looks good even 20–25 years later. And that claim cannot be made for many automobiles.

212 Export

When the various coachbuilders started building "customer" Ferraris in the early Fifties, they specialized in 2-passenger convertible coupes, and 2-passenger closed coupes—some with notchback and some with fast-back styling. An occasional 2+2 was built, most of them being dull and uninteresting, if not downright ugly. But this was 10 years before Ferrari introduced a 2+2 as part of his production series.

Carrozzeria Touring, Ghia, Ghia-Aigle (Swiss), Stabilimenti Farina, Bertone, Zagato and Abarth all built early Ferrari bodies, but the main contractors were first Touring, then Vignale, Boano, and finally Pininfarina and Scaglietti.

All the coachbuilders supplying bodies for Ferraris eventually settled on the now familiar eggcrate grille, but before that point—which occurred probably in 1953—a lot of experimentation was done. Not all of it successful.

Vignale went through a period in 1952 when most of his Ferrari bodies (designed by Giovanni Michelotti) carried grilles with a main horizontal and vertical cross-bar theme with a delicate vertical grillework set behind the protruding cross shape.

The vertical bar, furthermore, blended into a sharply defined center peak which extended the length of the hood. In our featured car, the front parking lights are incorporated into the grille assembly and the only thing that keeps this busy design from being overbearing is the lack of bumpers and other front end ornamentation. Not that bumpers are ornamentation, but they do add to a stylist's problems.

Other Michelotti/Vignale designs on this general theme were less successful because the addition of parking lights in the body—between grille and headlights—and rather bulky American-style

bumpers, created a surplus of both chrome plate and design components. One 340 America by Vignale has been seen with a 1940 Oldsmobile bumper on the front and, other than for its size, it looked surprisingly good. Given a simpler grille treatment the front end would have been exceptionally good and the car could have looked as though the bumper had been designed for it.

It seems highly incongruous, and maybe a bit unfair, to call a Ferrari "cute." But if any Ferrari ever qualified for that description, our featured 212 Export would have to be it. To some viewers, it appears to be feminine rather than masculine—as is expected of Ferraris.

Set on a short wheelbase, and with very little overhang in any direction, the car does look squat, compact and, in some eyes, bulbous. But a close examination of the shape, proportion and surface development reveals a handsome little car with rather pretty lines.

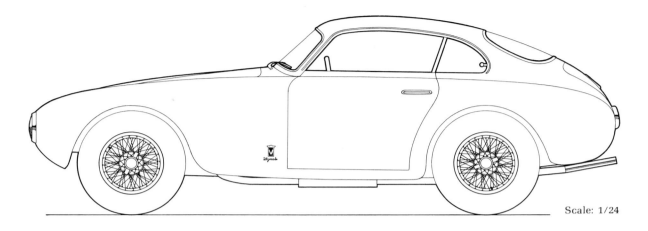

Scale: 1/24

212 Export

ENGINE

Type .... Colombo-designed, water-cooled, 60 degree V-12
Bore/stroke, mm/inches 68.0/58.8, 2.68/2.315
Displacement, cc/cubic inches 2562/156.3
Valve operation: Single overhead camshaft on each bank with finger
   followers & rocker arms to inclined valves
Compression ratio 8.0:1
Carburetion .... Three Weber 36 DCF twin choke, downdraft
BHP (Mfg) .... 170 @ 6500 rpm

DRIVE TRAIN

Clutch .... Single dry plate
Transmission: Five speed, non-synchromesh, direct drive in 4th gear
Rear axle .... Live
Axle ratio 4.66:1

CHASSIS

Frame .... Welded tubular steel, ladder-type
Wheelbase, mm/inches 2250/88.6
Track, front, mm/inches 1270/49.8
   rear, mm/inches 1250/49.2
Front suspension: Unequal length A-arms, transverse leaf spring, anti-
   roll bar
Rear suspension .... Live axle, semi-elliptic springs
Shock absorbers .... Houdaille hydraulic lever-action
Brakes .... Hydraulic, aluminum drums with iron liners
Tire size .... 5.90 x 15
Wheels .... Borrani wire, center-lock knock-off

GENERAL

Length overall, inches 148.0
Width 61.4
Height 51.0
Curb weight 2090
Body builder .... Alfredo Vignale

Regardless, it is a Ferrari, and concealed beneath the aluminum bodywork, with its striking red and black lacquer, is the typical Ferrari chassis and engine of the day. The drive was still on the right and the gearbox was still non-synchromesh, as Ferrari had not yet opted for all-synchromesh transmissions (to come in 1953), and left hand drive for his road cars except in rare instances. There have been Ferraris built as early as 1951 with left hand drive, but it didn't become standard until late 1953 or early 1954.

Even though I've gotten quite varied opinions about the appearance of this car, my opinion is that it is one of Michelotti's more successful designs and Carrozzeria Vignale built the body to his design with deft precision and superb craftsmanship. Its current location is unknown, but if found would certainly be one of the more desirable Ferraris to own.

340 Mexico

The first participation of Ferraris in the Carrera Panamericana in 1951 resulted in a 1-2 finish with Taruffi/Chinetti and Ascari/Villoresi in 2.6 liter 212 Export coupes. The two cars were series-built (at least as much series as Ferrari had at the time) in contrast with cars built specifically for racing, the engines were smaller in displacement than most of their competitors', and to say the finish order made an impact on the automotive racing world is an understatement.

This was also about the time that Ferrari had seriously started exporting cars and the North American market was a potential goldmine if properly approached. Enzo Ferrari realized the impact of the Carrera on this market and decided to go all out for the 1952 race. He reasoned that this long-distance, high-speed event would need special equipment if other manufacturers were as interested in the race as he was, so he ordered development of four cars to be specifically built for the 1952 race.

There were three coupes; for Alberto Ascari/Giuseppe "Nino" Farina, Luigi Villoresi/Franco Cornacchia, and Luigi Chinetti/Jean Lucas; and a roadster for American Bill Spear. All were mechanically identical with Lampredi-designed V-12s (the 1951 cars had Colombo engines) and 340 America drive trains. The stronger rear axle and all-synchromesh 4-speed transmission of the 342 were considered, but it was apparently felt that the available rear axle gearing for the 340 was better suited to the high speed race, and possibly the narrower track of the 340 (which would allow a narrower body and less drag) was a factor as well.

The Lampredi engine design was very similar to that of the earlier Ferrari V-12s designed by Colombo but the cylinder barrels were screwed into the cylinder heads instead of pressed into the block as they were in the earlier engines. The cylinder heads had

separate ports for each cylinder (Colombo's were siamesed). Both designs had single overhead camshafts on each bank which opened the valves via rocker arms, but the Colombo engines had finger followers while the newer Lampredi design used roller followers.

Because of the small bore, and the need to install or remove the rod and piston assembly from the top of the cylinder, the Colombo engines used connecting rods with the big end split at a 40 degree angle. The Lampredi design split the rod crank journal end at 90 degrees. The differences between the two engines added up to the big engine being stronger, but also having more inertia because of its heavier internal weight.

The four 340 Mexicos built for the 1952 race had frames constructed for Ferrari by Gilberto Colombo's firm, Gilco Autotelai. The chassis consisted of smaller diameter main frame tubes than were usually used in Ferraris, with a superstructure built up at the cowl and included the windshield pillars and roof rails (for the coupes) as part of the frame. The uprights were connected by a truss of smaller tubes and although they were not space frames in the sense that the 300SL or Birdcage Maserati frames were, Ferrari called them tubular and this is reflected in the serial number designation "AT" for America Tubolare.

Ascari's Mexico crashed on the first leg of the race in 1952 when he hit loose gravel on a curve and ran into an embankment. Neither he nor Farina were injured, but the car was out of the race. Villoresi posted the fastest times on the second day, and on the first leg of the third day—from Mexico City to Leon—but between Leon and Durango the pinion bearing on the rear axle gave out and his fine run was finished. Chinetti's Mexico finally finished in third place overall behind the 300SLs of Kling and Lang.

Chinetti took his car back to Europe, entering it in the Mille

Miglia, the 12 hours of Pescara and the Reims 12-hour race. It came back to the U.S. in 1954 and has remained here ever since. The Ascari and Villoresi cars were sold to Allen Guiberson. The Ascari car subsequently was sold to A. V. Dayton of Tulsa, and Phil Hill drove the ex-Villoresi car in the 1953 Carrera with Richie Ginther as passenger. Phil lost it on the short leg between Puebla and Mexico City when the clutch pedal would not return during a downshift and left him with no power to the rear wheels entering a tight curve. The crash did not injure either occupant, and didn't hurt the car much, but was bad enough to put them out of the race.

In 1961 Phil, being interviewed by *Road & Track,* was asked what he thought of Richie as a passenger or co-driver. Phil said, "Richie was great—of course he was better before we went over the cliff in 1953."

All four of the 340 Mexicos constructed for the 1952 Carrera are in the United States at the present time and, for the record keepers, the coupe serial numbers are 0222AT, 0224AT (our featured car and third place finisher in the 1952 Carrera) and 0226AT while the roadster carries number 0228AT. Even serial numbers were given to competition cars while odd numbers went to the touring cars.

These four 340 Mexicos are some of the most striking cars ever built by any coachbuilder for Ferrari. The hood on the coupe, from windshield base (which is actually behind the midpoint of the wheelbase) to the lip of the radiator opening, measures 77.5 inches—one of the longest hoods ever seen on a Ferrari. As a comparison, it is 5.5 inches longer than a 250 Europa, which had eight inches more wheelbase.

Styled by Michelotti for Carrozzeria Vignale, the body design is, with no qualification, wild. The "blac" device (boundry layer air control) on each door looks more effective than it actually is. These

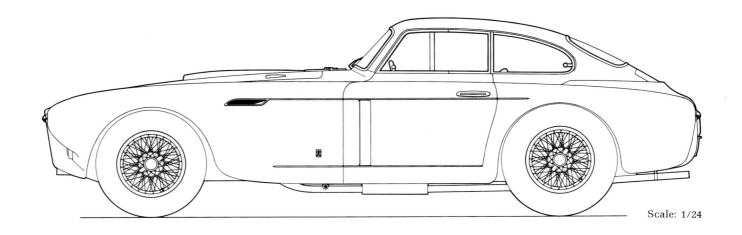

Scale: 1/24

340 Mexico

ENGINE

Type .... Lampredi-designed, water-cooled, 60 degree V-12
Bore/stroke, mm/inches 80.0/68.0, 3.156/2.68
Displacement, cc/cubic inches 4101/250
Valve operation: Single overhead camshaft on each bank with roller followers & rocker arms to inclined valves
Compression ratio...8.1:1
Carburetion .... Three Weber 40 DCF/3 twin-choke, downdraft
BHP (Mfg)...280 @ 6600 rpm

DRIVE TRAIN

Clutch .... Multiple disc
Transmission: Five speed, non-synchromesh, direct drive in 4th gear
Rear axle...Live
Axle ratio .... Optional—3.636, 4.000 & 4.444:1

CHASSIS

Frame: Welded small diameter steel tubing with rectangular boxed members
Wheelbase, mm/inches 2600/102.3
Track, front, mm/inches 1278/50.0
rear, mm/inches 1250/49.2
Front suspension .... Unequal length A-arms, transverse leaf spring
Rear suspension: Live axle, semi-elliptic springs and parallel trailing arms on each side.
Shock absorbers .... Houdaille hydraulic lever-action
Brakes .... Hydraulic, aluminum drums with iron liners
Tire size, front/rear .... 6.00/6.50 x 16
Wheels.................................. Borrani wire, center-lock knock-off

GENERAL

Length overall, inches 165.4
Width 59.2
Height 52.5
Curb weight 2200
Body builder .... Alfredo Vignale

were planned to channel boundry layer air at the side of the body into the rear brake air scoops at the back of the doors. In actual practice, they do nothing at all—aerodynamically—because they are in "dead air."

Nevertheless, and functional or not, these are exciting cars. The author has owned 0224AT for several years now and has driven it about 1,000 miles. It is cramped inside, steers like a truck, rides like a brick wagon, has dodgy brakes until those large drums are thoroughly warmed up, is hot in the summer and cold in the winter, but at the same time gives more personal pleasure per mile than just about any conveyance one could name. Looking out over that enormous hood length, one can't help but be impressed by it, and what is underneath it.

To get underway, you insert the key, turn one notch to the right, flip on the electric fuel pump, then hit the starter button and zowie! Instant bedlam. There is absolutely no insulation in this car anywhere, and every sound from the engine compartment is immediately part of the passenger compartment.

The clutch was a seven disc (four aluminum, three steel) multiple plate affair which has been converted to a five plate (three steel and two with lining on each side) assembly. The technique, however, for any Ferrari clutch is to let it in gently at low rpm *then* get on the throttle. One or two rev-it-to-5000-and-drop-the-clutch starts and the car would have no clutch left.

Using the accepted Ferrari technique, the car moves away quite briskly and once you've mastered the 5-speed non-synchro gearbox, fairly smooth changes can be made. The big Lampredi engine pulls reasonably well from anything over 1000 rpm, but if you wait until it's turning 3500 or so it really gives a belt in the back. I've never had my car over 6000 in any gear but if you're in fourth or fifth you're moving along at a pretty good rate of speed. And having become used to four-wheel disc brakes in recent years, at anything over a hundred—which is easy riding in a 340 Mexico—one feels just a bit insecure. I suppose, in 1952, most racing cars and most sports cars had somewhat similar stopping power, just as they would today, but driving a living museum piece, such as this car is, one is constantly reminded that he is driving a 1952 car in a mid-Seventies world.

250 Europa

The Paris auto show in October 1953 had one of the most important displays in the history of the seven year old Ferrari company. Sharing the Ferrari stand at the show were two new cars; the 250 Europa and 375 America. These were the largest Ferraris ever built at that time in terms of sheer bulk, and both were on the same 110-inch-wheelbase chassis.

In addition to sharing the same chassis, they both had the Lampredi long-block engine even though the Europa had only 2963cc compared to the 4522cc of the America. The reason for two engine sizes in one basic car was different market requirements. The European buyer faced stiff purchase taxes based on engine size, so the Europa 250 was made for that market. The American buyer, with no such limitations, could have the greater horsepower of the 375 if he wanted it, or could opt for the smaller engine.

In spite of the size of these cars, they were clean, handsome, and set a future pattern for automotive styling trends that would continue for years. Both bodies were by Pininfarina, and it was

Ferrari's first real effort to get into limited production with a series-built car.

Starting around serial number 0295 EU, about 20 long-engined 3 liter Europas were built—through number 0351 EU. A word here about Ferrari serial numbers; even numbers were given to competition cars and the occasional one-off or special, and odd serial numbers went to what would be normally called "road cars" or "touring cars." Unfortunately, cars with successive serial numbers were not necessarily built consecutively, which makes it difficult for the historian.

Our featured car is not one of those production units even though its serial number, 0313 EU, puts it in the middle of the series, but is a styling exercise by Carrozzeria Alfredo Vignale.

If the three previous Michelotti/Vignale designs in this book (195 I, 212 EX and 340 Mexico) are considered "successful" then this 250 Europa show car has to be considered one of their failures. Displayed first at the Geneva auto show in 1954, the design

appeared on the 250, then a few months later at the Turin show an almost identical body which, to a slight degree, came off better was shown on a 375 America chassis. These were the only two cars to be built with this body configuration.

What is basically a fairly decent shape is spoiled by garish over ornamentation. Our featured car has been further defiled by some California custom de-chroming, the removal of the original bumpers and the addition of "nerfing bar" bumperettes front and rear.

Show cars, usually, are opinion-testers for future models—at least in the U.S.—but European coachbuilders traditionally receive chassis from the major car builders for the creation of strictly one-off designs for the resultant publicity. If the design is successful, both manufacturer and coachbuilder benefit. If not, the coachbuilder loses because the chassis manufacturer shrugs it off by saying "we only supplied the chassis, we didn't design the car." Good or bad, show cars in Europe are often sold to customers at the end of the auto show, and this Europa found its way to California. The location of the 375 America is unknown.

As was typical with Michelotti/Vignale designs, all the various slots, louvers and scoops were functional in that they were actually openings in the bodywork. How well they functioned is questionable. We suspect that the air ducts on this car were not any better than the ones on the 340 Mexico and in this case were put there more for decoration than for function. Correct or not, the designers were honest enough to not want fake scoops on their cars.

It seems a pity that this car is so over-embellished because the basic form and proportions are not all that bad. Given less surface decoration, the car probably would look quite nice. It almost has the "Targa" look and the body shape would have lent itself to that treatment with excellent results. But, even the best of talents seems to have an occasional lapse.

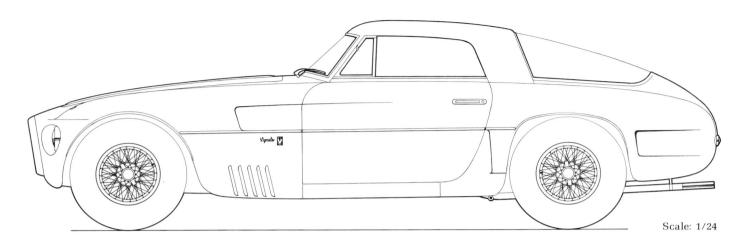

Scale: 1/24

250 Europa

ENGINE

Type .... Lampredi-designed, water-cooled, 60 degree V-12
Bore/stroke, mm/inches 68.0/68.0, 2.68/2.68
Displacement, cc/cubic inches 2963/181
Valve operation: Single overhead camshaft on each bank with roller
    followers & rocker arms to inclined valves
Compression ratio 8.5:1
Carburetion .... Three Weber 36 DCZ twin-choke, downdraft
BHP (Mfg) .... 200 @ 7000 rpm

DRIVE TRAIN

Clutch .... Single dry plate
Transmission: Four speed, all synchromesh, direct drive in 4th gear
Rear axle .... Live
Axle ratio 4.25:1

CHASSIS

Frame .... Welded tubular steel, ladder-type
Wheelbase, mm/inches 2800/110
Track, front, mm/inches 1325/52.2
    rear, mm/inches 1320/52.0
Front suspension .... Unequal length A-arms, transverse leaf spring
Rear suspension: Live axle, semi-elliptic springs and parallel trailing
    arms on each side
Shock absorbers .... Houdaille hydraulic lever-action
Brakes .... Hydraulic, aluminum drums with iron liners
Tire size .... 7.10 x 15
Wheels .... Borrani wire, center-lock knock-off

GENERAL

Length overall, inches 173.5
Width 63.0
Height 54.5
Curb weight 2800
Body builder .... Alfredo Vignale

Show cars, such as this one, seldom had modifications to the engine or chassis as they were built strictly for show—and in the case of Ferrari they had quite a bit of natural "go" so there wasn't really any need for further mechanical modifications.

These first long wheelbase Europas still used the now traditional Ferrari front suspension with unequal length A-arms and transverse leaf spring. About the same time the 375 car was being shown at Turin, though, the second series Europa (which carried the GT suffix after the serial number instead of EU) was introduced on a 102.5 inch wheelbase, and was the first Ferrari to have coil springs in the front suspension.

The live rear axle with semi-elliptic springs and twin parallel trailing arms were retained, and by now the 4-speed, all-synchromesh gearbox was pretty much standard, as was left hand steering.

The extra long wheelbase was needed for the long-block engine and a desire to have plenty of legroom in the passenger compartment. The engine was heavy and the layout caused severe understeer so the first Europas were not as pleasant to drive as were the later, second-series Europas with six inches less wheelbase and the short, lighter Colombo engine. The long wheelbase cars were at home on superhighways anywhere, but in cities or in the mountains they were a handful to drive. And with no power steering they were definitely a man's car.

Aside from the extra length, and the fact that it had a small-displacement big engine, the Europa was similar in almost every respect to other Ferraris of the time. The company had settled on a basic chassis layout which was adhered to with only minor variations from model to model until much later in the company's history when independent rear suspension and transaxles came along on front engined cars, and still later when some of the road cars would have mid-engines.

375 Mille Miglia

Through the 1952 season, Ferrari had built and raced (in sports and GT) cars with 1497, 1902, 1995, 2341, 2562, 2715 and 2953cc capacity Colombo-designed V-12 engines; and 3322 and 4101cc Lampredi-designed V-12s. Competition from Jaguar, Maserati and Lancia was getting tougher, however, so for the 1953 Le Mans 24-hour race in June Ferrari entered two 4101cc 340 MM berlinettas and a 4494cc 375 MM berlinetta—all with Pininfarina bodywork.

In outward appearance, the cars resembled the 250 MM Pininfarina coupes, but the 375 had a slightly longer wheelbase than the 340s. The race was not altogether successful for Ferrari; one 340 MM, driven by Gianni and Paolo Marzotto finished 5th, the Hawthorn/Farina 340 was disqualified for taking on oil before it had completed the required 22 laps before the first pit stop, and the new 375 MM driven by Alberto Ascari and Gigi Villoresi retired with clutch failure after 12 hours of exchanging the lead with both the Rolt/Hamilton and Moss/Walker Jaguars. Ascari did set the lap record for the race at 112.85 mph.

At the 24 hour Touring Car GP at Spa-Francorchamps circuit in Belgium on July 25–26, 1953, a team of three 4.5 liter 375s were entered for Maglioli/Carini (retired with clutch trouble), Ascari/Villoresi (retired with a broken rear axle) and Farina and Hawthorn who won the race at 94.84 mph average. This was the first victory for a 375 Ferrari.

Before the Belgian event, the large rear windows seen at Le Mans had been replaced with smaller windows installed in metal panels fitted into the large former rear window openings. This was apparently to reduce headlight glare from just-passed (or, heaven forbid, overtaking) cars in the nighttime hours of these long distance races for which the 375 was designed, and seemed well suited.

Also new on the 375 coupes for the Spa race was a lower nose with headlights faired in under clear plastic which was contoured to the fender lines. They were fearsome looking beasts.

Gianni Marzotto won the Senigallia race on August 9 in a low-nosed 375 MM, and a week later on August 15, Mike Hawthorn and Umberto Maglioli teamed up to win the Pescara 12-hour race at 79.18 mph in a similar car.

On August 30, three Ferraris were sent to the Nurburgring 1000 Km race—a 375 MM coupe for Hawthorn/Villoresi, a 375 MM open roadster (with Vignale bodywork) for Ascari/Farina, and a 4-cylinder, 3-liter car for Maglioli/Carini. All three cars practiced but only the Ascari/Farina roadster actually started the event, winning the race after the two leading Lancias dropped out. (This 375 roadster is the car that Phil Hill/Richie Ginther drove to second place in the Carrera Panamericana in 1954.)

The 1953 Carrera Panamericana drew a full team of Ferraris, all 375 MMs, entered by Franco Cornacchia's Scuderia Guastalla. Luigi Chinetti was in an open roadster with Vignale body, Guido Mancini, Mario Ricci and Antonio Stagnoli were in the low-nosed Pininfarina berlinettas as raced since the July Spa race, and Umberto Maglioli was in a newly-designed Pininfarina-bodied berlinetta.

Backing up these quasi-factory entries were private entrants Phil Hill (in Guiberson's 1952 340 Mexico berlinetta) and Ruiz Echeverria in a 250 MM berlinetta.

By now the Mexican race meant a lot to Ferrari but this was not his year to win it. On the first leg, between the Tuxtla-Gutierrez start and Tehuantepec, Stagnoli's car blew a tire at full speed (about 170 mph). The subsequent crash killed his co-driver Scotuzzi immediately and Stagnoli died later in the hospital. On the fifth stage Maglioli's car lost a rear wheel at over 125 mph and, while the occupants were uninjured, the car was out of the race. When the accident occurred, Maglioli and Fangio were vying for the lead

The factory berlinettas at Le Mans in 1953 looked like larger versions of the 250 MM. This is the Ascari/Villoresi car which retired with clutch trouble while in second place after having set the fastest lap time of 4 min, 27.4 sec.

with no more than a minute between them after 10 hours of racing.

At this point Maglioli took over Ricci's car which was in eighth place, 2 hours, 32 minutes behind Fangio's leading Lancia. Maglioli won the last three legs by virtue of all-out superb driving but could only reduce Fangio's lead by 27 minutes, finishing sixth overall.

Fortunately for Ferrari, Mancini's 375 was fourth which gave Ferrari enough points to barely win the 1953 manufacturer's championship over Jaguar. Echeverria's 250 MM was seventh; Chinetti was retired on the second leg, by disqualification, after his car couldn't finish the leg within the allotted time; and Phil Hill's 340 went out on the third leg when the clutch pedal wouldn't return leaving Phil stranded with no power to the rear wheels as he entered a curve. Phil and passenger Richie Ginther were unhurt but the car was out of the race.

The first year of competition for the 375 MM was not total success, nor was it total disaster. And the racing world knew that 1954 was going to be a tough year if you weren't driving a Ferrari.

The primary purpose of a racing berlinetta was to compete in long distance, high speed races where aerodynamics and driver comfort would be more important than lighter weight and subsequent maneuverability. Thus, the Ferrari roadsters (or spyders, as they were referred to by the factory), were most often seen in closed circuit, tight-quarter racing where acceleration and handling were all important.

The berlinettas, which weren't really much heavier than the open cars, were usually entered in the Mille Miglia, Le Mans, Spa-Francorchamps, Reims or Mexico. Quite obviously, there were overlaps both ways as the factory did a lot of experimenting with engine sizes and body shapes, and customers usually had a preference for an open or closed car to suit their own whims.

Mario Ricci (23) near Tehauntepec on the first day of the 1953 Carrera, and Guido Mancini (26) near Oaxaca on the second day—both in 375 MMs. Mancini finished fourth, behind the Lancias of Fangio, Taruffi and Castellotti which gave Ferrari the 1953 Manufacturer's Championship. Below, the long-block Lampredi engine which was the basis for the 250 Europa, 340 and 342 America, 340 Mexico, 375 America, 375 Mille Miglia and 375-Plus engines.

Umberto Maglioli's new 375 MM just north of Oaxaca on the second day of the 1953 Carrera Panamericana. This car later lost a rear wheel and Maglioli took over Ricci's 375 MM to finish the race. Maglioli's car is seen again, with three variants of the same design by Studio Pininfarina—all built in 1953.

This Pininfarina-designed 375 MM, shown at the Turin auto show in 1955, set the basic style for future berlinettas.

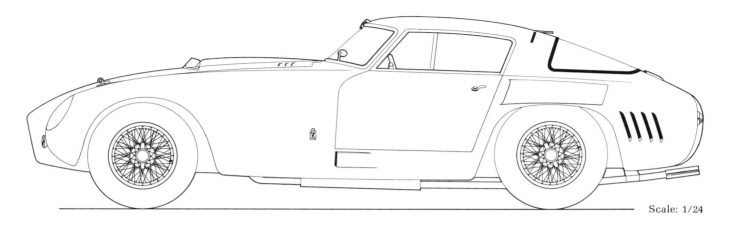

Scale: 1/24

375 Mille Miglia

ENGINE

Type .... Lampredi-designed, water-cooled, 60 degree V-12
Bore/stroke, mm/inches 84.0/68.0, 3.307/2.68
Displacement, cc/cubic inches 4522/275.8
Valve operation: Single overhead camshaft on each bank with roller followers & rocker arms to inclined valves
Compression ratio..9.2:1
Carburetion ....Three Weber 40IF4C four-choke, downdraft
BHP (Mfg)...340 @ 7000 rpm

DRIVE TRAIN

Clutch .... Multiple disc
Transmission: Four speed, all-synchromesh, direct drive in 4th gear
Rear axle..Live
Axle ratio .... Optional—3.44, 3.56, 4.00, 4.43:1

CHASSIS

Frame .... Welded tubular steel, ladder-type
Wheelbase, mm/inches 2600/102.3
Track, front, mm/inches 1325/52.2
   rear, mm/inches 1320/52.0
Front suspension: Unequal length A-arms, transverse leaf spring and anti-roll bar
Rear suspension: Live axle, semi-elliptic springs and parallel trailing arms on each side
Shock absorbers .... Houdaille hydraulic lever-action
Brakes .... Hydraulic, aluminum drums with iron liners
Tire size, front/rear .... 6.00/7.00 x 16
Wheels.............................. Borrani wire, center-lock knock-off

GENERAL

Length overall, inches 170.0
Width 60.0
Height 52.0
Curb weight 2400
Body builder .... Pininfarina

Carrozzeria Pininfarina started building Ferrari bodies in 1952, and in 1953 was the sole constructor of racing berlinettas for the factory. Touring and Stabilimenti Farina were no longer building bodies for the factory. Vignale had started to construct bodies for road-going touring Ferraris to be sold to customers, and continued to build sports racing roadsters.

According to official Ferrari figures there were only 93 Ferraris built during 1953; including touring, grand touring, sports racing and Formula 1 and 2 single-seat racing cars. We have no breakdown of how many of each was built, and it would certainly be interesting to know how many survive today.

Our featured car, no. 0322 AM is one of those raced at the 1953 Carrera Panamericana and very likely competed at Le Mans and Spa-Francorchamps earlier in the year as well. Dick Irish drove the car to seventh place in the main event at March Air Force Base in 1954. At some point after that the small rear window was replaced, by a larger one similar in size and shape to the ones first seen on the cars at Le Mans in 1953, probably in the interest of driver visibility.

The chassis of the 375 MM was based on the 342 America, including the 4-speed, all-synchromesh gearbox and the heavier

center section for the live rear axle. Rear springs were semi-elliptic, but location was by double trailing arms on each side, which also served to absorb acceleration and braking torque.

Front suspension was still independent with unequal length A-arms on either side, and a transverse leaf spring. Shock absorbers were Houdaille lever type all around.

The engines used in the 375 MM berlinettas were the Lampredi design developed for Formula 1 Grand Prix racing. Typical of Ferrari engines, the bore was larger than the stroke at 84 x 68 mm for a total displacement of 4522cc. Ignition was by twin Marelli magnetos at the back of the engine, and carburetion was by three four-throat Weber IFC/4 carburetors. The resultant horsepower was 340 at 7000 rpm.

Having referred to the 340 Mexico as "striking," "wild" and "exciting" to look at, the 375 MM can only be classed as "brutal," "mean" and "aggressive" looking. Standing beside one of these machines gives the viewer the impression it could pounce at any moment and devour anything in its way, man or man-made.

And the 375 was not a car that one drove to the race. The horsepower and torque generated by the big V-12 was brutal, and the car's response to that power—through the multiple disc racing clutch—was instantaneous and left no room for inept driver reaction.

With 340 bhp on tap in a car that weighed about 2400 lbs at the curb, one could get in a lot of trouble at any speed if not careful. A further problem of Ferraris of this period was the weight distribution change between full and empty fuel tanks. The 375 MM (also the 250 and 340) carried a 40 gallon tank, which meant a weight difference of nearly 300 lbs between full and empty, behind the rear axle.

These cars were, and are, not for the timid or faint of heart, They required muscle, stamina, and full attention to every detail of driving to survive, let alone win a race.

at four points (rubber insulated) and a fifth at the back of the gearbox.

In spite of the 250 MM's theoretical potential, it never achieved great success in international competition for some reason. The record books show a few 1953 victories—and they don't differentiate between spyder and berlinetta—such as Villoresi first and Farina third at the Autodrome GP at Monza; Paolo Marzotto first in the Dolomite Gold Cup at Belluna; and in 1954 Trintignant and Piotti were first in the Hyeres 12-hour race.

The 250 MM, both coupe and roadster, was considerably more successful in private hands in races throughout the world—Phil Hill in particular won a number of races on the west coast in his Vignale-bodied 250 MM spyder.

Cars, like people, are unpredictable, and cars which had far less theoretical potential often did better than others that seemingly should be unbeatable.

In 1953–54 Ferrari's main competition came from Maserati, Jaguar and Aston Martin. Mercedes-Benz had retired the first 300SLs and the 300SLRs didn't compete until 1955. Cunningham and Talbot-Lago were, of course, forces to be reckoned with at Le Mans but neither competed on as vast a scale as did Ferrari so Talbots were seldom seen outside France, and Cunninghams were raced mainly in France and the U.S.

Racing record or not, the Pininfarina 250 MM berlinetta is one of the most sought-after Ferraris and one that is commanding higher and higher prices—assuming one can be found that is for sale. The combination of style, size (it is only 157 inches long and 49 inches high) and performance make it a most desirable car to own. With its aluminum body and lack of bumpers it is unlikely that an owner would drive one of these delightful machines as everyday transportation but, mechanically, it could be (and has been) done.

Mackay Frazer, an American who had just started to drive for BRM before his death in 1957 drove a 250 MM berlinetta for a road car, and Roger Ellis of Reno, Nevada, had over 100,000 miles on his 250 MM when he sold it in 1974.

but with its shorter wheelbase, and better balance, it was far more maneuverable on short, twisty circuits. Lighter weight undoubtedly was also a factor in its agility as the 250 berlinetta weighed in the neighborhood of 2000 lbs compared to the 2200–2400 of the 340 and 375 MMs.

A significant difference between the 250 S/MM of Bracco's and the production 250 MM was the switch from 5-speed non-synchromesh transmission to a 4-speed all-synchro box. Otherwise, the chassis were the same, and carried on the then current Ferrari tradition; independent front suspension with unequal length A-arms, and transverse leaf spring located by four Silentbloc bushings.

Rear suspension was by two semi-elliptic springs and twin parallel trailing arms on each side to locate the live rear axle, and take acceleration and braking torque. Damping was by Houdaille hydraulic lever-action shock absorbers all around.

As in previous models, the frame was a welded up assembly of large diameter tubing with a tubular X brace in the center, fabricated steel box member at the front and large diameter tubing cross members at the rear. The engine was mounted in the chassis

500 Mondial

The young Ferrari organization was continually experimenting with engines and chassis, often switching Grand Prix and sports car components, to achieve exactly the right configuration for various types of competition.

Late in 1951, Aurelio Lampredi began development of a four cylinder engine, in two sizes, for Ferrari to use in Formula 1 and Formula 2 races. The larger of the two, at 2.5 liters, had a bore and stroke of 94 x 90mm, and the 2-liter unit had a bore and stroke of 90 x 78mm.

Ferrari's reasoning for the development of a four cylinder engine was that the low-speed torque characteristics of the four cylinder design would result in better performance on short, twisty circuits (compared to his high-revving V-12s). Furthermore, Ferrari felt that the emphasis would be on Formula 2 racing, which was limited to two liters displacement, in 1952. He was correct in both assumptions and Alberto Ascari won the World Championship for drivers in 1952 and 1953, driving the four cylinder, 2-liter Ferrari. During the two seasons, Ascari won 11 out of 15 races, and won nine in a row.

Continuing his cross-pollination engineering, the 2-liter engine was also being developed into a sports car engine in 1953, and was installed in a prototype car that was to be called the Mondial. The prototype chassis design was similar to the 250 MM, including the transmission location at the engine, but before it was ready for sale to the public considerable changes were made to the chassis.

The Mondial, as first sold to customers, had the original Ferrari-type front suspension with two unequal length A-arms on each side and a transverse leaf spring retained by Silentbloc bushings. But the rear suspension was now De Dion. The De Dion axle beam is located by twin, parallel trailing arms on each side and the spring is a transverse leaf.

Also unlike the prototype Mondial, the customer car had a 4-speed transmission at the rear in unit with the axle center section which contained a self-locking differential.

A second-series Mondial was built later, which had coil spring front suspension and a cylinder head from the Type 553 Formula 2 engine. All the first series Mondials had Pininfarina bodies, but the second series cars were bodied by Scaglietti to a design by Enzo Ferrari's son, Dino. But we're ahead of our story.

Sometime in 1953 two berlinettas were built, one by Pininfarina and one by Vignale, on 500 Mondial chassis. No record of competition is known for either car, and while the Pininfarina version is in the United States, the location of the Vignale coupe is unknown.

Both were built on the series 1 Mondial chassis and therefore have the 1984.8cc four cylinder, twin-cam engine producing 160 bhp at 7000 rpm (the series 2 cars had 170 bhp at 7000). Dry sump

lubrication is used on the engine with a double-bodied pressure/scavenge pump and a 16 liter separate oil tank reservoir.

Twin Marelli magnetos supply the spark to two plugs per cylinder, and carburetion is by two Weber 40 DCOA 3, twin-choke, sidedraft carburetors.

Drive is taken through a multiple disc clutch to the rear-mounted 4-speed transmission. An oil pump and filter are incorporated in the gearbox. Again, a change made in the series 2 Mondials was the use of a double, dry plate, friction-lined clutch and a 5-speed transmission.

With the proliferation of engine, chassis and body types made by Ferrari in the early Fifties, it is surprising that only two four cylinder coupes seem to have been built.

For customers who had no plans to race, but wanted the Ferrari name on their cars and that wonderful 12 cylinder noise, there obviously was no choice. But after Ascari's back-to-back championships in '52 and '53 in four cylinder, 2-liter Ferraris, it would seem that racing oriented customers would have purchased more four cylinder cars.

Unfortunately the Mondial never became a real winner and as

Scale: 1/24

500 Mondial

ENGINE

Type .... Lampredi-designed, water-cooled, inline four
Bore/stroke, mm/inches 90.0/78.8, 3.531/3.062
Displacement, cc/cubic inches 1984.8/121
Valve operation: Double overhead camshafts with tappet followers
   operating directly on inclined valves
Compression ratio...8.5:1
Carburetion .... Two Weber 40 DCOA3 twin-choke, sidedraft
BHP (Mfg) ....160 @ 7000 rpm

DRIVE TRAIN

Clutch .... Multiple disc
Transmission: Four speed, all-synchromesh, direct drive in 4th gear
Rear axle... De Dion
Axle ratio .... Optional—3.92, 4.12, 4.34, 4.55, 5.06:1

CHASSIS

Frame .... Welded tubular steel, ladder-type
Wheelbase, mm/inches 2250/88.6
Track, front, mm/inches 1278/50.0
   rear, mm/inches 1284/50.6
Front suspension .... Unequal length A-arms, transverse leaf springs
Rear suspension: De Dion axle beam, transverse leaf spring and parallel
   trailing arms on each side
Shock absorbers .... Houdaille hydraulic lever-action
Brakes .... Hydraulic, aluminum drums with iron liners
Tire size, front/rear .... 5.50/6.00 x 16
Wheels................................ Borrani wire, center-lock knock-off

GENERAL

Length overall, inches 147.0
Width 60.0
Height 49.0
Curb weight 1750
Body builder .... Pininfarina

Lampredi had left Ferrari to go to work for Fiat, Ing. Massimino was called in to develop the 500 Testa Rossa as a replacement for the Mondial.

Our Pininfarina-bodied feature car has a more than casual resemblance to PF designs on larger Ferrari chassis, notably the 250 MM and the 375 MM driven by Maglioli in the 1953 Carrera Panamericana. It also shares, to some extent, the shape of the open roadsters with Pininfarina bodywork. Built on the 88.6 inch-wheelbase Mondial chassis, it isn't quite as well proportioned as the larger cars, but it is a handsome little car. With a total length of 147 inches, and a height of 49 inches, it is one of the smallest Ferrari road cars ever built and at 1750 lbs it is one of the lightest.

I asked the current owner if he knew the torque rating of the engine and he said no, but it's "like a John Deere." That's not surprising when you consider that each cylinder of the Mondial's engine has 500 cc and anyone who's ridden a well-tuned, high output 500 cc single-cylinder motorcycle knows about torque.

When the name Ferrari is mentioned, the first thought is of V-12 engines at high revs, emitting that high-pitched scream which is a part of their character. However, Ferrari has built inline four, V-6, and V-12 road cars; and inline four, inline and V-6, flat opposed eight and V-8, and both flat opposed and V-12 engines for his race cars.

The two Mondial 4-cylinder berlinettas pictured are, as far as is known, the only four cylinder coupes ever built on Ferrari chassis. They therefore have to be considered as two of the rarest cars ever built, and certainly two of the most desirable from a collector's standpoint.

250 GT

International sports car racing had progressed so far and so fast after World War II that by 1955, the cars were virtually GP cars with two-passenger enclosed bodywork. This was particularly true of the Mercedes 300SLR and the Ferrari 121 LM roadsters.

Then after Pierre Levegh's disastrous accident at Le Mans in 1955, there was a clamor to return to the classic type of racing with dual purpose sports cars. Specific classes were established by the F.I.A. for *gran turismo* cars starting with the 1956 season.

Ferrari had built berlinettas since 1950 and these had been raced by both factory and customers against open roadsters. In long distance races such as the Mille Miglia, Le Mans, Reims, Spa and the Carrera Panamericana, the closed cars had done well and established a credible racing record overall.

So when new regulations were established for GT cars, Ferrari was not completely without a basis for creating a new car to compete successfully in these classes.

Coincidentally, Pininfarina had built bodies on several 250 GTs in 1955 that looked like variations of the 375 MM (which in turn had been derived from the 250 MM) bodies from the cowl back, but had sloping hoods leading to elliptical-shaped radiator openings.

One car was shown at the Paris show in 1955, which had a beautiful shape but suffered from over adornment in the form of chrome trim and unfortunate fins on the rear fenders, all of which was uncharacteristic of Studio Farina. At the Turin show a few weeks later another version appeared with sliding plexiglas windows and a centrally located quick-fill gas cap at the rear.

At the Geneva show in March, 1956 the Farina stand displayed one of the best all around designs yet seen on a Ferrari; a derivative of the 1955 Paris and Turin cars, it stood out because it was so clean and uncluttered. These Pininfarina designs in 1955–56 were the forerunners of the Scaglietti berlinettas as raced by Ferrari and sold to customers for the 1956 GT season.

Scaglietti, in Modena, was now the official builder of Ferrari berlinetta coachwork and even though faithfully following the original Farina shape, Scaglietti did soften the form somewhat, causing it to be more voluptuous.

In the year prior to the establishment of the new GT classes, Ferrari had been involved with Grand Prix racing and, to a lesser extent, with the major events in the Manufacturer's Championship, all but ignoring all lesser events and the existing GT races. As a result, Armando Zampiero won the Italian Sports Car Championship in 1955 driving a Mercedes 300SL.

Ferrari therefore lost no time in getting his new 250 GT berlinettas into the hands of customers for the upcoming season.

The first victory for this new series of Ferraris was won by the Marques Alfonso de Portago at Nassau in December, 1955.

Early in 1956 the Belgian Olivier Gendebien with his cousin, Jacques Wascher, drove a 250 GT berlinetta to fourth overall and first in class in the Tour of Sicily, then at the end of April almost duplicated that performance by finishing fifth overall and first in GT class in the Mille Miglia.

The Tour de France is a timed rally around the perimeter of the country. The fifth Tour, held in September, 1956, was 3600 miles long, including six races held on major circuits (Comminges, Le Mans, Rouen, Reims, St Etienne, and Montlhery); two hill climbs (Mont Ventoux and Peyresoude) and a drag race (at Aix-les-Bains). The team of De Portago and Nelson won five of the six circuit races and gained first overall in the Tour.

Throughout 1956 the 250 GT berlinettas continued to rack up victories and some of the private entrants had an astounding season. Edoardo Lualdi Gabardi (often racing under the name "Lualdi") won five 1st places and three 2nds; Giuliano Giovanardi had five 1sts, two 3rds and an 8th place; and Camillo Luglio, who had an even lighter Zagato body on his 250 GT, won five 1sts and two 2nds.

Scaglietti began construction in January of 1957 on a new series of berlinettas similar in shape to the 1956 cars but the cold-air scoop on the hood became standard, the large rear windows were replaced by smaller windows and louvered quarter panels (behind

The 1958 250 GT berlinettas had the headlights covered by faired-in clear plastic, three large air outlets replacing the row of 14 louvers in the quarter panel, and squared-off rear fenders with the taillights set right at the fender tips.

Gendebien/Wascher Scaglietti-bodied 250 GT, in which they won the GT class, at the start of the 1956 Mille Miglia. Below, the engine of the 1959 250 GT.

the side window) were added. Once again Scaglietti stayed close to the Farina design, but softened the general shape a bit. If anything, these 1957 cars have to be considered better looking than the first version and, in this writer's opinion, better than the subsequent designs to come in 1958–59.

The headlights of these 1956 and 1957 GT berlinettas were mounted at the front of the fender but recessed about one inch into chrome bezels (California customizers call this "frenching" the lights).

The 1957 season dispelled any doubts about the superiority of the Ferrari 250 GT berlinetta. Starting out on April 14, Gendebien/Wascher won the Tour of Sicily outright, with the 250 GTs of Luglio and Lena finishing 5th and 6th. On May 12, Gendebien/Wascher finished third overall in the Mille Miglia with an average speed of 93.6 mph. Piero Taruffi's dohc 4-liter Ferrari spyder averaged only 94.8 in winning the race, and it was reported that on some parts of the course the 4-cammers had difficulty getting past the Belgians' 250 GT.

On July 13–14, 250 GTs finished 1st, 2nd, 3rd, 4th and 5th in the 12 hours of Reims, the winner being Gendebien/Frere at 104 mph. They also set the lap record for the race at 115 mph.

The 1959 Scaglietti 250 GT berlinetta reverted to uncovered headlights.

the cylinder heads are of the original Colombo siamesed-port design, and the magnitude of Ferrari development can be appreciated when one realizes the bhp obtained from the 1958–59 version of the very first Colombo design for Ferrari in 1946.

A four speed, all-synchromesh transmission is coupled to the V-12 via a single plate dry clutch. Wheelbase is 102.3 inches and the overall length is 173.0 inches. The weight is kept down to 2520 by the use of aluminum for the body panels.

Although built for competition, these 250 GT berlinettas make ideal touring cars if one could stand the possibility of damage to the relatively fragile aluminum body. Careless door opening in parking lots which does minor damage to a steel body, will do far more to an aluminum body, and with only minimal bumpers the front and rear are virtually unprotected as well.

If the owner is willing to risk this aggravation the 1956–1959 long wheelbase berlinettas are some of the most tractable and most enjoyable Ferraris to drive of all the now accepted "classic" Ferrari designs.

250 GT

In 1959, three distinct berlinetta designs appeared; all Farina-designed, Scaglietti-built. The first was a variation of the 1958 design, but with the headlights now uncovered, with chrome rims, so it looked more like a standard road car than a competition car.

The second design was a complete departure from the 1956–58 style with rounded contours everywhere contrasting with the more pronounced and sometimes sharp changes of contour on the earlier cars. The hood came down low between the front fenders to a tucked-under grille. The headlights were set in the front of the fenders in chrome bezels and were the foremost part of the bodywork.

The rear overhang was short but graceful and the passenger compartment was notable for its large glass area. Instead of the "traditional" quarter panel with some variation of louvered air outlets it now had small quarter windows, and door vent windows.

This design was built on the 102.3-inch wheelbase of the 1958 cars and there was no significant change in the chassis or engine. The body design presaged what was to come later in the year on a shorter (94.5 inch) wheelbase. The car is generally referred to as an interim 250 GT, although the factory continued its standard 250 GT designation.

These interim cars were built in very limited quantity, probably no more than a total of six cars. But while the chassis and engine were as seen before on GT berlinettas, a French auto magazine published a cutaway drawing of an interim GT showing a Testa Rossa-type 12-port engine with six twin-choke Webers and sparkplugs located on the outside of the heads (the Colombo-designed engines normally had the plugs on the inside).

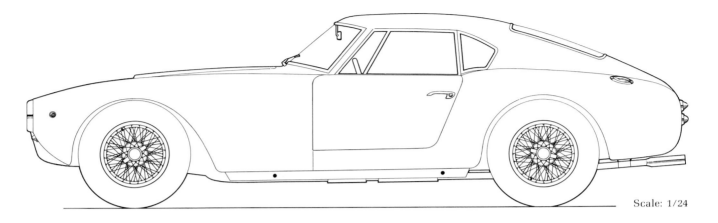

Scale: 1/24

250 GT

ENGINE

Type ....Colombo-designed, water-cooled, 60 degree V-12
Bore/stroke, mm/inches 73.0/58.8, 2.87/2.315
Displacement, cc/cubic inches 2953/180.0
Valve operation: Single overhead camshaft on each bank with roller
    followers & rocker arms to inclined valves
Compression ratio...9.57:1
Carburetion ....Three Weber DCL3 twin-choke, downdraft
BHP (Mfg)...260 @ 7000 rpm

DRIVE TRAIN

Clutch ....Single dry plate
Transmission: Four speed, all-synchromesh, direct drive in 4th gear
Rear axle...Live
Axle ratio ....Optional—3.67, 3.78, 4.00, 4.25, 4.57, 4.86:1

CHASSIS

Frame...............................Welded tubular steel, ladder-type
Wheelbase, mm/inches 2600/102.3
Track, front, mm/inches 1354/53.3
    rear, mm/inches 1349/53.1
Front suspension...........................Unequal length A-arms, coil springs
Rear suspension: Live axle, semi-elliptic springs and parallel trailing
    arms on each side
Shock absorbers ....Houdaille hydraulic lever-action
Brakes ....Hydraulic, aluminum drums with iron liners
Tire size, front/rear .... 6.00 x 16
Wheels...........................Borrani wire, center-lock knock-off

GENERAL

Length overall, inches 173.0
Width.. 65.0
Height 50.5
Curb weight 2520
Body builder .... Scaglietti

Gendebien's 250 GT interim car during a race at Spa-Francorchamps which was part of the 1959 Tour de France. Gendebien was the Tour winner.

The cutaway also showed disc brakes on the car. This was not the standard engine for the car however, nor were disc brakes standard. The first racing appearance of the interim model was at Le Mans in 1959 where Pilette/Arents and Fayen/Munaron placed 4th and 6th overall. In September the team of Gendebien/Bianchi won their third-in-row Tour de France in another interim 250 GT, which was the 4th consecutive Tour victory for Ferrari. It is not known by this writer if the Le Mans or Tour de France cars had the Testa Rossa engine, but it seems likely that the Le Mans cars, at least, did, because of their overall high placing.

The interim design has never been as popular with Ferrari enthusiasts as either the preceding or subsequent cars for some reason. Some tend to consider the cars "neither fish nor fowl" because of the mixture of long wheelbase chassis and short wheelbase bodywork. This is unfortunate because they really are significant cars, particularly from the coachbuilder's standpoint as they did preview the upcoming 250 GTs from Scaglietti.

One recent owner in Los Angeles used his interim 250 every day (if his wife wasn't driving it) to commute to work, and reported it to be more than satisfactory in all respects. It was docile, tractable and yet had substantial performance if desired. All in all, it was and is a desirable car.

Inter-Europa, also in a swb berlinetta.

The victories for the new berlinetta continued through 1960. They were 1st (Mairesse/Berger), 2nd (Schlesser/Loustel) and 3rd (Tavano/Martin) in the Tour de France, and Gendebien and Bianchi won the 1000 km of Montlhery with their swb in October.

Only the major victories have been given here for the 1960 season, but the berlinettas in the hands of private entrants were doing extremely well all over Europe. Enzo Ferrari has credited Giotto Bizzarrini for a large part of the success of the 250 GT in 1960, and Ferrari wisely didn't change his winning combination for 1961.

Some minor differences could be seen in the bodies. After the 1959 show debut, side air outlets had been added to front and rear fenders, and for 1961 a vent window in the door was added. The top contour of the door and side window seemed smoother, and added a tiny bit of improvement to an already good looking car.

The first International event for 1961 was the Sebring 12-hour race on March 25. Motor sports writer Denise McCluggage and Jazz saxophonist Alan Eager placed their swb berlinetta in 10th overall and 1st in GT. At Snetterton, England, on the same day, Mike Parkes and Graham Whitehead were 1st and 2nd in their swb berlinettas ahead of Salvadori's Aston Martin DB4/GT.

Jaguar had announced the new XK-E on March 15, 1961 and racing fans were expecting big things from this exciting new car. The scheduled XK-E debut at the Fordwater Trophy Race at Goodwood, England on April 3rd didn't materialize and Mike Parkes and Graham Whitehead were 1st and 4th in swb berlinettas with the Ireland and Moss Aston Martins sandwiched inbetween.

The E-Type made an outstanding debut when it did appear by placing 1st and 3rd (Graham Hill and Roy Salvadori) at Oulton Park, England on April 15—one month after its announcement to the public. Innes Ireland was 2nd in an Aston Martin GT and the best the Ferraris could do was 4th and 5th with the Sears and Whitehead swb berlinettas. Ferrari, it seemed, now had some real competition.

Two weeks later, at Spa-Francorchamps, Willy Mairesse spoiled the XK-E's Continental debut by placing his Ferrari swb first. Mike Parkes had switched to Jaguar and finished 2nd, with Whitehead 3rd in his 250 swb.

The Le Mans trials, which are traditionally held two months before the race in June, demonstrated the swb 250 GT's potential when the Parkes/Tavano car lapped consistently at 120 mph. Unfortunately, it was destroyed by Jo Schlesser, but had proven faster than any of the sports/racing cars at the trials with the exception of the sports Ferraris and the Birdcage Maseratis.

In the actual 24-hour race, Noblet/Guichet and Grossman/Pilette finished 3rd and 6th overall for 1st and 2nd in GT class with their swb berlinettas.

The victories for Ferrari and his short wheelbase berlinettas went on through 1961. The Rob Walker swb, driven again by Stirling Moss, had placed as high as 3rd overall at Le Mans before a fan blade came off, cutting the lower water hose and causing the engine to overheat. This rare, right hand drive, car was returned to Modena for an engine rebuild and Moss won the British Empire Trophy Race (for GT cars) at Silverstone on July 8, 1961. The next day, Willy Mairesse drove his swb to first place in the Six Hours of Auvergne.

Moss won again, in Rob Walker's swb, at the Bank Holiday International Meeting at Brands Hatch on August 7, and finished first in the 26th RAC Tourist Trophy at Goodwood, August 19, with Mike Parkes 2nd in a normal left-hand drive swb berlinetta. Parkes won at Snetterton, on October 3, Pedro and Ricardo Rodriguez won the 1000 km of Montlhery on October 22, and Stirling Moss was triumphant at Nassau in December; all in swb berlinettas.

The Tour de France had become a Ferrari benefit it seemed. A Ferrari won the initial Tour in 1951, placed 2nd in both 1952 and

Scale: 1/24

250 GT

ENGINE

Type ....Colombo-designed, water-cooled, 60 degree V-12
Bore/stroke, mm/inches 73.0/58.8, 2.87/2.315
Displacement, cc/cubic inches 2953/180.0
Valve operation: Single overhead camshaft on each bank with roller
    followers & rocker arms to inclined valves
Compression ratio..9.2:1
Carburetion ....Three Weber DCL3 twin-choke, downdraft
BHP (Mfg) ....280 @ 7000 rpm

DRIVE TRAIN

Clutch ....Single dry plate
Transmission: Four speed, all-synchromesh, direct drive in 4th gear
Rear axle..Live
Axle ratio ....Optional—3.44, 3.55, 3.67, 3.78, 4.00, 4.25, 4.57:1

CHASSIS

Frame................................Welded tubular steel, ladder-type
Wheelbase, mm/inches 2400/94.5
Track, front, mm/inches 1354/53.3
    rear, mm/inches 1349/53.1
Front suspension: Unequal length A-arms, coil springs and anti-roll
    bar
Rear suspension: Live axle, semi-elliptic springs and parallel trailing
    arms on each side
Shock absorbers ....Houdaille hydraulic lever-action
Brakes ....disc
Tire size, front/rear ....6.00 x 16
Wheels................................Borrani wire, center-lock knock-off

GENERAL

Length overall, inches 163.5
Width 65.0
Height 50.5
Curb weight 2380
Body builder ....Scaglietti

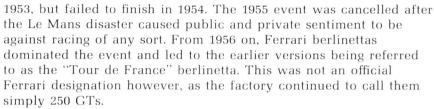

1953, but failed to finish in 1954. The 1955 event was cancelled after the Le Mans disaster caused public and private sentiment to be against racing of any sort. From 1956 on, Ferrari berlinettas dominated the event and led to the earlier versions being referred to as the "Tour de France" berlinetta. This was not an official Ferrari designation however, as the factory continued to call them simply 250 GTs.

The Marques Alfonso de Portago won the Tour in 1956, Ferrari GT berlinettas were 1st (Gendebien/Bianchi), 2nd (Trintignant/Picard) and 3rd (Lucas/Malle) in 1957; Gendebien and Bianchi were 1st again in 1958 with Da Silva Ramos 3rd; the Gendebien/Bianchi team won their third straight Tour in 1959 with an interim model, Mairesse/Berger were 2nd in a 1958 lwb GT, and De Lageneste/Schild finished third in an interim model.

The 1960 Tour de France was another 1—2—3 victory for Ferrari swb berlinettas, in the hands of Mairesse/Berger, Schlesser/Loustel and Tavano/Martin. In September 1961 Ferrari laid claim to "ownership" of the Tour de France by finishing 1st, 2nd, 3rd and 4th (Mairesse/Berger, Gendebien/Bianchi, Trintignant/Cavrois and Berney/Gretener).

The combination of good looks, superb performance, and an outstanding competition record firmly establishes the Ferrari 250 GT berlinetta—in all its forms—as one of the most successful and desirable cars ever built. They competed successfully in rallies, races, hillclimbs and concours d'elegance, neither giving nor asking quarter from any competitor. To many, it is the epitome of Ferrari automobiles.

250 GT

The talented and extremely prolific Carrozzeria Bertone organization, headed by Nuccio Bertone, built very few bodies for Ferraris, but the few Bertone did build were outstanding. The best record we have indicates Bertone built an attractive Cabriolet on a 166 Inter chassis in 1950, and a handsome coupe on a 250 GT chassis which was shown at the 1960 Geneva auto show (both cars are illustrated in the Fitzgerald/Merritt Ferrari book).

Their third, and final, effort for Ferrari during these formative years was also the best from this highly respected concern. The design was by Bertone's chief stylist Giorgio Giugiaro and reflected an influence from the Ferrari Formula 1 GP single-seaters, and the sports/racing cars, which had twin "nostril" air intakes for the radiators during the 1961 and 1962 seasons.

This last Bertone design was shown at the Geneva auto show in March, 1962. After the show, the car was purchased by an Italian industrialist who used it very little on the road as he had bought it for display purposes.

As shown at Geneva, the car was painted dark blue, and had a chrome prancing horse emblem on the underside of the nose between the grilles. Later, the prancing horse was moved from the center to the left grille opening, a row of small louvers was added behind each quarter window, the Bertone escutcheon (below the front fender side vents) was re-designed, and the car was painted light grey-green. The newer, lighter color, in this writer's opinion, shows the car off better than did the dark blue.

The Giugiaro/Bertone design was built on a 250 GT short wheelbase chassis, as detailed in the previous chapter, and it retained all the excellent mechanical qualities of that configuration

plus a unique body shape. It was a one-off body, and has never been duplicated.

Because of its steel body and added creature comforts such as roll-up windows, sound proofing, more deluxe upholstery, etc. this car weighs more than the competition car it was derived from. Performance is therefore diminished a bit, but is still outstanding, and the overall combination of its plusher interior and racing heritage created an exceptional road car.

By 1962, when this car was built, Mercedes and Lancia had full independent suspension. Ferrari had tried De Dion rear axles on some competition cars, and had gone to full independent suspension on his mid-engined racing cars. The normal Ferrari (if any Ferrari can be called *normal*) competition and road cars, however, still used live rear axles with semi-elliptic springs, located by twin trailing arms on each side.

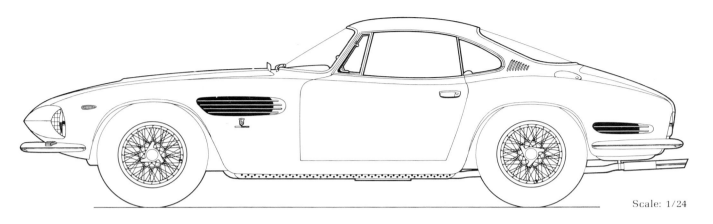

Scale: 1/24

250 GT

ENGINE

Type ....Colombo-designed, water-cooled, 60 degree V-12
Bore/stroke, mm/inches 73.0/58.8, 2.87/2.315
Displacement, cc/cubic inches 2953/180.0
Valve operation: Single overhead camshaft on each bank with roller followers & rocker arms to inclined valves
Compression ratio 8.8:1
Carburetion ....Three Weber DCL3 twin-choke, downdraft
BHP (Mfg) ....260 @ 7000 rpm

DRIVE TRAIN

Clutch .... Single dry plate
Transmission: Four speed, all-synchromesh, direct drive in 4th gear and Laycock de Normanville overdrive
Rear axle .... Live
Axle ratio 4.57:1

CHASSIS

Frame .... Welded tubular steel, ladder-type
Wheelbase, mm/inches 2400/94.5
Track, front, mm/inches 1354/53.3
   rear, mm/inches 1349/53.1
Front suspension: Unequal length A-arms, coil springs and anti-roll bar
Rear suspension: Live axle, semi-elliptic springs and parallel trailing arms on each side
Shock absorbers .... Houdaille hydraulic
Brakes .... disc
Tire size .... 6.00 x 16
Wheels .... Borrani wire, center-lock knock-off

GENERAL

Length overall, inches 165.0
Width 65.0
Height 55.0
Curb weight 2700
Body builder .... Carrozzeria Bertone

Ferrari had many critics of his apparent reluctance to give up the rear live axle in favor of independent rear suspension and the GT cars retained live axles through the GTO series. But there are also those who believe that a well-developed live axle is preferrable to an unproven independent design. Regardless, Ferrari continued to win races with his "archaic" designs and when the time came that full independent suspension seemed necessary, the switch was made—on the 275 GTB series.

In 1962, when our Salon car was built, Ferrari had almost 14 years of development on this basic chassis layout and it worked extremely well as anyone who's driven one of these cars can verify. It was good for its day, and is still good on any but the worst of roads. The 250 GT Bertone Ferrari's current owner has driven it almost 75,000 miles and has hopes of driving it another 75 or more—at least he has no plans to dispose of the car. I know, I inquired about its "availability."

Other Cars

Ghia-bodied 340 America, equipped with Halibrand mag wheels, finished fifth in the 1952 Carrera Panamericana, driven by Jack and Ernie McAfee.

Kurt Zeller's 166 MM berlinetta, built in 1953, carried an attractive Pininfarina body.

Similar configuration can be seen on this Ghia-built
166 Inter. Egg-crate grille was not yet standard.

Franco Cornacchia's Vignale-bodied 195 Sport,
in the Coppa Inter-Europa race in 1950.

Bracco's 250 S which won the 1952 Mille Miglia,
and led the Carrera Panamericana, was the
forerunner of the 250 MM berlinetta.

The 1951 Mille Miglia winner was this
Vignale-bodied 340 America, driven by
Villoresi and Cassani.

This 1955 prototype competition berlinetta
is similar to the Turin show car at right.

Vignale-bodied 212 Export was driven by Phil Hill and
Arnold Stubbs to sixth place in 1952 Carrera Panamericana.

In 1952, Vignale was experimenting with fins and
two-tone paint schemes. Some worked, and some didn't.

Zagato "double bubble" top was designed to achieve
low frontal area, but leave headroom for passengers.
This is Camillo Luglio at the start of the Mille Miglia.

A 250 GT with 375 MM type bodywork was shown
by Pininfarina at the 1955 Turin auto show.

Pininfarina's design for this 375 MM
berlinetta, shown at Turin in 1955, set
the theme for future Ferraris.

Scaglietti's first Ferrari berlinetta was
this 1955 design for Roberto Rossellini
on a 375 MM chassis.

Engine	166 Mille Miglia	195 Inter	212 Export	340 Mexico	250 Europa
Designer	Colombo	Colombo	Colombo	Lampredi	Lampredi
Type	V-12	V-12	V-12	V-12	V-12
Bore & Stroke, mm	60.0/58.8	65.0/58.8	68.0/58.8	80.0/68.0	68.0/68.0
Bore & Stroke, in	2.362/2.315	2.562/2.315	2.68/2.315	3.156/2.68	2.68/2.68
Displacement, cc/in	1995/122	2340/143	2562/156.3	4101/250	2963/181
Compression Ratio	8.5:1	7.5:1	8.0:1	8.1:1	8.5:1
Camshaft Layout	S.O./Bank	S.O./Bank	S.O./Bank	S.O./Bank	S.O./Bank
Camshaft Drive	Chain	Chain	Chain	Chain	Chain
Cam Followers	Finger	Finger	Finger	Roller	Roller
Valves—Design, No	Inclined, 2/Cyl	Inclined, 2/Cyl	Inclined, 2/Cyl	Inclined, 2/Cyl	Inclined, 2/Cyl
Valve Springs	Hairpin	Hairpin	Hairpin	Hairpin	Hairpin
Spark Plugs/Cyl	1	1	1	1	1
Ignition	2 Magnetos	2 Distributors	2 Distributors	2 Distributors	2 Distributors
Carburetors—No, Type	3-36 DCF	1-32 DCF	3-36 DCF	3-40 DCF/3	3-36 DCZ
BHP/RPM	125/7000	130/6000	170/6500	280/6600	200/7000

Drive Train

Clutch	Single Dry Plate	Single Dry Plate	Single Dry Plate	Multiple Disc	Single Dry Plate
Transmission—Gears	5-Speed & Rev	5-Speed & Rev	5-Speed & Rev	5-Speed & Rev	4-Speed & Rev
Type	Non-Synchro	Non-Synchro	Non-Synchro	Non-Synchro	All-Synchro
Location	With Engine	With Engine	With Engine	With Engine	With Engine
Axle Ratios	4.66 or 5.00:1	4.66 or 5.00:1	4.66 or 5.00:1	3.636, 4.000, 4.444:1	4.25:1

Note: All specifications are for cars featured in this book and while other body styles of the same series will often be identical mechanically, some dimensions may be different—particularly overall length, height, and weight.

375 Mille Miglia	250 Mille Miglia	500 Mondial	250 GT (LWB)	250 GT (SWB)	250 GT
Lampredi	Colombo	Lampredi	Colombo	Colombo	Colombo
V-12	V-12	Inline 4	V-12	V-12	V-12
84.0/68.0	73.0/58.8	90.0/78.0	73.0/58.8	73.0/58.8	73.0/58.8
3.307/2.68	2.87/2.315	3.531/3.062	2.87/2.315	2.87/2.315	2.87/2.315
4522/275.8	2953/180	1984.8/121	2953/180	2953/180	2953/180
9.2:1	9.0:1	8.5:1	9.57:1	9.2:1	8.8:1
S.O./Bank	S.O./Bank	D.O.H.C.	S.O./Bank	S.O./Bank	S.O./Bank
Chain	Chain	Gears	Chain	Chain	Chain
Roller	Roller	Tappet	Roller	Roller	Roller
Inclined, 2/Cyl	Inclined, 2/Cyl	Inclined, 2/Cyl	Inclined, 2/Cyl	Inclined, 2/Cyl	Inclined, 2/Cyl
Hairpin	Hairpin	Hairpin	Hairpin	Coil	Coil
1	1	2	1	1	1
2 Magnetos	2 Magnetos	2 Magnetos	2 Distributors	2 Distributors	2 Distributors
3-40 IF4C	3-36 IF4C	2-40 DCOA3	3-DCL3	3-DCL3	3-DCL3
340/7000	240/7200	160/7000	260/7000	280/7000	280/7000

Multiple Disc	Multiple Disc	Multiple Disc	Single Dry Plate	Single Dry Plate	Single Dry Plate
4-Speed & Rev	4-Speed & Rev	4-Speed & Rev	4-Speed & Rev	4-Speed & Rev	4-Speed & Rev
All-Synchro	All-Synchro	All-Synchro	All-Synchro	All-Synchro	All-Synchro
With Engine	With Engine	With Rear Axle	With Engine	With Engine	With Engine
3.56 or 4.000:1	3.636, 4.000, 4.444:1	3.92, 4.12, 4.34, 4.55, 5.06:1	3.67, 3.78, 4.00, 4.25, 4.57, 4.86:1	3.44, 3.55, 3.67, 3.78, 4.00, 4.25, 4.57:1	4.57:1

	166 Mille Miglia	195 Inter	212 Export	340 Mexico	250 Europa
Chassis	Welded Tube	Welded Tube	Welded Tube	Welded Tube	Welded Tube
Frame	Ladder-Type	Ladder-Type	Ladder-Type	Semi Truss	Ladder-Type
Wheelbase—mm/in	2250/88.6	2500/98.4	2250/88.6	2600/102.3	2800/110
Track, Fr—mm/in	1270/49.8	1270/49.8	1270/49.8	1278/50.0	1325/52.2
Track, Rear—mm/in	1250/49.2	1250/49.2	1250/49.2	1250/49.2	1320/52
Suspension—Front	Independent	Independent	Independent	Independent	Independent
Type	Un. Length A-arms	Un. Length A-arms	Un. Length A-arms	Un. Length A-arms	Un. Length A-arms
Springs	Trans. Leaf	Trans. Leaf	Trans. Leaf	Trans. Leaf	Trans. Leaf
Suspension—Rear	Rigid Axle	Rigid Axle	Rigid Axle	Rigid Axle	Rigid Axle
Springs	Semi elliptic	Semi Elliptic	Semi Elliptic	Semi Elliptic	Semi Elliptic
Shock Absorbers	Houdaille	Houdaille	Houdaille	Houdaille	Houdaille
Brakes	Drum	Drum	Drum	Drum	Drum
Wheels	Borrani, Wire	Borrani, Wire	Borrani, Wire	Borrani, Wire	Borrani, Wire
Tires—Fr/Rear	5.90 x 15	5.90 x 15	5.90 x 15	6.00/6.50 x 16	7.10 x 15

General

	166 Mille Miglia	195 Inter	212 Export	340 Mexico	250 Europa
Body Designer			Michelotti	Michelotti	Farina
Body Builder	Touring	Touring	Vignale	Vignale	Pininfarina
Length Overall—in	146.0	156.0	148.0	165.4	173.5
Width	60.0	60.0	61.4	59.2	63.0
Height	46.0	47.0	51.0	52.5	54.5
Curb Weight	2100	2150	2090	2200	3080

375 Mille Miglia	250 Mille Miglia	500 Mondial	250 GT (LWB)	250 GT (SWB)	250 GT
Welded Tube	Welded Tube	Welded Tube	Welded Tube	Welded Tube	Welded Tube
Ladder-Type	Ladder-Type	Ladder-Type	Ladder-Type	Ladder-Type	Ladder-Type
2600/102.3	2400/94.5	2250/88.6	2600/102.3	2400/94.5	2400/94.5
1325/52.2	1300/51.1	1278/50.0	1354/53.3	1354/53.3	1354/53.3
1320/52	1320/52	1284/50.6	1349/53.1	1349/53.1	1349/53.1
Independent	Independent	Independent	Independent	Independent	Independent
Un. Length A-arms	Un. Length A-arms	Un. Length A-arms	Un. Length A-arms	Un. Length A-arms	Un. Length A-arms
Trans. Leaf	Trans. Leaf	Trans. Leaf	Coils	Coils	Coils
Rigid Axle	Rigid Axle	De Dion	Rigid Axle	Rigid Axle	Rigid Axle
Semi Elliptic	Semi Elliptic	Trans. Leaf	Semi Elliptic	Semi Elliptic	Semi Elliptic
Houdaille	Houdaille	Houdaille	Houdaille	Houdaille	Houdaille
Drum	Drum	Drum	Drum	Disc	Disc
Borrani, Wire	Borrani, Wire	Borrani, Wire	Borrani, Wire	Borrani, Wire	Borrani, Wire
6.00/7.00 x 16	5.50/6.00 x 16	5.50/6.00 x 16	6.00 x 16	6.00 x 16	6.00 x 16

Farina	Farina	Farina	Farina	Farina	Giugiaro
Pininfarina	Pininfarina	Pininfarina	Scaglietti	Scaglietti	Bertone
170.0	157.0	147.0	173.0	163.5	165.0
60.0	63.0	60.0	65.0	65.0	65.0
52.0	49.5	49.0	50.5	50.5	55.0
2400	2080	1750	2520	2380	2700

AFTERWARD

If you bought this book and have read this far, the odds are that you have a more than casual interest in Ferraris. Welcome to a large—and getting larger—band of enthusiasts.

As a 2-time Ferrari owner I've been called a masochist (because I'll put up with the myriad problems Ferraris are supposed to have), rich (because of the cost of buying and maintaining a Ferrari), stupid (for spending that much when I could have gotten a better car for less money—this usually from Porsche or Corvette owners), or lucky (because I own one—this from hopeful someday Ferrari owners), and many variations of the above.

The simple facts, as I see them anyway, are that a Ferrari is neither as good as some of its boosters would have you believe, or as bad as its detractors would have you believe. But good or bad, a Ferrari is not an indifferent car, and there are far too many of those available to drivers who just don't care how they get from point A to B. I do care.

I drove my first Ferrari daily and the only trouble I had with it was from a non-Ferrari part; a previous owner had installed one of those after-market electronic ignitions and it never worked right. When we got rid of that, everything was fine.

My present Ferrari, the 340 Mexico, is not driven daily, but it isn't a fear of breakdown that keeps it in the garage. The car is over 22 years old, has aluminum bodywork with no bumpers, was built for racing and is therefore not at its best in commuting traffic, and it is worth a lot of money. I want to enjoy it as long as I can, then see it end up in a good museum or in the hands of another enthusiast who will enjoy it as much as I have. But it does get driven.

There are those, however, who do use their Ferraris, albeit newer, more civilized ones, every day with no more trouble than they would have from any other make of car. One acquaintance, who is with a large national real estate chain, and drives a lot in his business, put over 200,000 miles on a 1964 250 GT Lusso with no serious problems of any kind.

One of the worst types of driving for any car is the short-haul start it up and shut it off routine where the working parts never get warmed up. And this is particularly bad for cars with aluminum or magnesium castings in engine, transmission or differential.

When I drove my car daily, I started it up each morning and as soon as the oil was circulating I held the throttle steady with the engine running at about 2000 rpm—never blipping the throttle on a cold engine as you often see and hear in the racing pits, which is a "grandstanding" ego-trip for the ignorant driver or mechanic. As soon as the water temperature gage needle started to move off the peg, I backed out and drove away, keeping the engine under 2500 in any gear for the first few miles so that engine, gearbox and rear axle could warm up before being driven harder. Once the parts are all at working temperatures, you can get on it.

A Ferrari is not unbreakable, but if the driver will allow time for all the working parts to get up to working design temperatures, it is hard to break a Ferrari. They are meant to be driven hard and fast and it can be done with little fear of breakage. *But Ferraris must be warmed up correctly and sufficiently before being driven hard.*

A Ferrari mechanic once told me the biggest problem he had with any Ferrari was its owner. Too many owners are either the jump-in-the-car-and-put-their-foot-in-it kind or they insist on tinkering with the engine when they don't know the first thing about it.

A good mechanic (not a parts changer) is not awed by the fact that it has 12 cylinders, or that the camshafts are in the heads instead of in the block, or that it might have two distributors instead of one. If he has metric tools, and a thorough grounding in automobile mechanics, he can probably do most of the maintenance on a Ferrari without too much trouble. For any major work I'd still take it to a specialist.

Ferrari lore is not helped by some of the salesmen who work either in a "Ferrari agency" or on a used car lot that happens to acquire one for resale. Too many of these characters are supercilious jerks who know far too little about either automobiles or human nature. The misinformation, or out-and-out lies I've heard from salesmen with a Ferrari to sell is incredible.

Another pet peeve of mine, admittedly not a critical one however, is the misspelling and mispronunciation of names associated with Ferrari. I don't expect everyone to know how to pronounce foreign names, many have no reason to have learned, but for people in the business (who should know better) to perpetrate the error seems wrong somehow.

So, as a final section in this book, and a small tribute to those who helped make this book possible, we present a very short discourse on names of coachbuilders, drivers, and others closely associated with Ferrari—and particularly Ferrari berlinettas. It may not help you to know more about Ferraris, but you'll sound as though you do, and a good bit of one upsmanship often comes in handy.

COACHBUILDERS & DESIGNERS

Bertone	sounds	like	Bear-tony
Carrozzeria Touring	"	"	Ka-rote-zer-eeya Touring
Ghia	"	"	Ge-ah (hard g as in gone)
Giovanni Michelotti	"	"	Gee-oh-von-ne Mick-e-lotty
Pininfarina	"	"	Pee-neen-fa-ree-na
Scaglietti	"	"	Skahl-yetty (g is silent)
Stabilimenti Farina	"	"	Stah-bil-e-menty Fa-ree-na
Superleggera	"	"	Super-ledg-er-ah
Alfredo Vignale	"	"	Al-fray-doh Vin-yahl-e

DRIVERS & DESIGNERS

Carlo Mario Abate	sounds	like	Kar-loh- Mar-ee-o Ah-botty
Alberto Ascari	"	"	Al-bear-toh As-car-ey
Luigi Bazzi	"	"	Loo-ee-gee Bot-zey
Giovanni Bracco	"	"	Gee-oh-vonny Brock-oh
Luigi Chinetti	"	"	Loo-ee-gee Kin-etty
Gioacchino Colombo	"	"	Gee-oh-ah-keeno Co-lom-bo
Franco Cornacchia	"	"	Fraun-co Kor-nock-ee-ah
Giuseppe Farina	"	"	Gee-oo-seppe Fa-ree-na
Olivier Gendebien	"	"	Oh-liv-ee-ay Zhan-de-bee-en
Jean Guichet	"	"	Zhan Ge-shay (hard g)
Vittorio Jano	"	"	Vee-tor-ee-o Yawn-o
Aurelio Lampredi	"	"	Aw-rayl-ee-o Lam-pray-de
Umberto Maglioli	"	"	Um-bear-toh Mahl-yoh-le
Willy Mairesse	"	"	Willy My-ress
Guido Mancini	"	"	Gwee-doh Man-cheeney
Alberto Massimino	"	"	Al-bear-toh Mass-e-mee-no
Mario Ricci	"	"	Mar-ee-o Ree-chee
Giorgio Scarlatti	"	"	Ge-orge-ee-o Scar-lotty
Antonio Stagnoli	"	"	Awn-tony-o Stahn-yoh-le
Piero Taruffi	"	"	Pee-air-oh Ta-rroof-ey
Maurice Trintignant	"	"	Maw-reece Trin-tin-yawn

CIRCUITS

Mille Miglia	sounds	like	Mill Mill-ya
Monza	"	"	Mone-za
Piacenza	"	"	Pee-a-chen-za

BOOK DESIGN by Chuck Queener.

ACKNOWLEDGEMENTS

All books are the result of the combined efforts
of many persons. Even though I am both author
and publisher, this book could not have been
published without the expert help of the following:
COVER COLOR SEPARATIONS—Roberts Graphic Arts, El Monte, Calif.
TYPESETTING—Computer Typesetting Services, Glendale, Calif.
PRINTING—Rotary Offset Printers, Anaheim, Calif.
RESEARCH INFORMATION (publications)—*Ferrari The Sports
& GT Cars* by Fitzgerald & Merritt, *Ferrari* by Hans Tanner,
The Ferrari V-12 Sports Cars 1946-56 by Anthony Pritchard,
*Road & Track, Sports Cars Illustrated, Autocourse, Automobile
Year* and *Sports Cars of The World.* (individuals)—Roger
Ellis, John Iglehart, Bill Karp, Fred Leydorf, Ed Niles,
John Queen, Chuck Queener, Hans Tanner and Jon Thompson.

PHOTO CREDITS

Gino Barbieri, Dean Batchelor, Bob Behme, Bertone, Bernard
Cahier, Alex Callier, Jerry Chesebrough, Ghia, Geoff Goddard,
John Iglehart, Image International, Russ Kelly, Strother
MacMinn, Henry Manney, Corrado Millanta, Rodolfo Mailander,
George Paxton, Ralph Poole, Pininfarina, Publifoto, Lee
Render, Joe Rusz, Scaglietti, Dennis Shattuck, Vignale,
Kurt Worner, Zagato.

Scale drawings by Dean Batchelor and Bob Price